GROSS
SCIENCE
EXPERIMENTS

GROSS SCIENCE EXPERIMENTS

60 SMELLY, SCARY, SILLY
TESTS TO DISGUST YOUR FRIENDS AND FAMILY

EMMA VANSTONE

Author of *This Is Rocket Science* and *Snackable Science Experiments*

PAGE STREET
PUBLISHING CO.

PAGE STREET
PUBLISHING CO.

First published in 2020 by
Page Street Publishing Co.
27 Congress Street, Suite 105
Salem, MA 01970
www.pagestreetpublishing.com

Distributed by Macmillan, sales in Canada by The Canadian Manda Group.

24 23 22 21 20   1 2 3 4 5

ISBN-13: 978-1-64567-114-5
ISBN-10: 1-64567-114-3

Library of Congress Control Number: 2019957304

Cover and book design by Molly Gillespie for Page Street Publishing Co.
Photography by Emma Vanstone
Emoji illustrations © Shutterstock/Carboxylase, illustration on page 19 © Shutterstock/Kraphix

Printed and bound in China

DEDICATION

For Zak, Sydney, Hannah and Charlie xxx

CONTENTS

INTRODUCTION

Get ready to shriek and squeal all the way through this bone-chillingly gruesome book. Make and pick fake scabs, create a blood bath, drink a blood cocktail and dissect a brain. Test your surgical skills by cutting open a fake stomach, and try to identify healthy urine through taste . . . yuck!

If you're after something a bit more lighthearted, try making fake vomit, sticky poo or a bug-filled bath bomb. Find out why saliva is super and how it impacts taste, then create the most revolting food combination you can imagine.

If smells are more your thing, there are eggs, farts, stinky wee and even cow burps in the Stinky Smells chapter (page 47) and all kinds of poo in the All About the Poo chapter (page 109).

Learn all about bugs by making a worm hotel, a bug restaurant and a real slime trail, or take a journey back in time to experiment with mummification, plastic milk and different kinds of toilet paper.

Find out about disgusting diseases and the grossest things humans have done throughout history with these deliciously horrible experiments.

You might get a bit sticky, slimy and dirty, but that's the best kind of science, right? Many of the projects in the book are open-ended and have ideas for further exploration on the topic. Don't be limited by the ideas in the book either. Tinkering, improving designs and generally having fun with materials are what's most important.

Emma Vanstone

HOW TO USE THIS BOOK SAFELY

1. Follow the instructions! Read through the instructions first, and if you're not sure about something or need to use a heat source, ask an adult for help. If you're the adult reading this, choose projects you think your child can do safely alone, or plan to work together with them on the more difficult ones.

2. Don't eat or drink any experiments unless the activity says you'll be creating something edible, and if you're out and about, never assume any poo is the delicious edible kind that's on page 69.

3. Look after your eyes and clothes. Remember, food coloring can stain clothes and furniture, and slime can be tricky to remove too!

4. Look after animals. I promise no animals were harmed in the making of this book, and I expect the same consideration from YOU! Remember that worms and bugs are super helpful to our planet and we NEED them!

5. Always ask an adult before starting an experiment.

6. The final product—where there is one—doesn't have to look perfect, and it's totally fine to make a mistake along the way. Who knows, you might end up with an even better creation or investigation.

7. Don't forget to clean up after you're finished, especially if you've made a mess. Your parents will thank you!

HOW TO THINK LIKE A SCIENTIST

People tend to use the word *experiment* in a very general way, but a real experiment has a specific format that follows the scientific method.

You need to ask questions, come up with a hypothesis, or prediction, and set up an investigation, usually changing only one variable. When a real scientist has finished experimenting, they analyze the results and form a conclusion. Phew! It sounds complicated but really isn't. Check out the Toilet Paper Testing experiment on page 141 to find out more.

The best way to conduct science activities is to watch what happens, predict what will happen next, observe what happens next and then explain why it happened.

ASK QUESTIONS

If you're a child, ask your adult or helper questions. You might be able to teach them something new.

If you're the adult, asking simple questions about the activity you've just done is a great way to see if the child understands what has happened. Encourage them to ask questions too, as a question can often lead to a whole new activity.

EXTEND THE LEARNING

Once you've finished an activity, the learning can be extended in lots of fun ways:

- **MAKE AN ANNOTATED DRAWING**
  Draw what happened and label it. Showing the drawing to friends and family can be a great opportunity to share your knowledge with others.

- **WRITE YOUR OWN INSTRUCTIONS**
  If you've found a new way to approach an activity or improve on the instructions, write down the new instructions and ask a friend to follow them.

- **TURN IT INTO A STORY**
  "Once upon a time, two children walking in the woods came across a frozen gelatin brain in the water. It didn't look real. Was it a trick?"

You get the idea!

BLOOD
AND BRAINS

Ancient Egyptians used brains to cure eye infections and pulled them out through the nose with a hook. Aristotle thought the brain was for cooling the blood, and Einstein's brain was smaller than average. The brain is a truly incredible organ that stores memories and controls how we think and react to the environment around us. The human brain contains billions of nerve cells and weighs around 3 pounds (1.4 kg).

Blood makes up around 7 percent of the weight of a human body and contains red blood cells, white blood cells and platelets. In the past, people thought losing blood was beneficial to health, so they would purposely remove blood from people in an attempt to cure their ailments. Leeches were sometimes used to do this, as they can suck several times their own body weight in blood! We now know that humans generally don't survive if they lose 40 percent of their blood volume, or about 4 to 5 pints (2 to 2.5 L).

In this chapter, you'll make sticky scabs to pick, create fake blood to shock your friends, drink a deliciously gross blood cocktail, prepare a blood bath and dissect a brain.

PICK A SCAB

As soon as you cut your skin, platelets jump to the rescue. They stick together around the damaged area to form a clot. Clots are thick and sticky, made up of blood cells and fibrin. This clot becomes what we call a scab, which stops nasty microorganisms and toxins from infiltrating the opening in your skin.

It might feel quite satisfying to pick a scab, but underneath there's a lot going on with dead bacteria, live bacteria, white blood cells and maybe even a bit of pus. Yuck! Picking that scab doesn't sound so appealing now, does it?

Scabs usually start off red in color, darken as the scab dries and thickens and then become lighter as the skin is repaired. Signs of infection such as yellow pus should be examined by a doctor. Otherwise, it's really important to leave scabs alone so the damaged cells underneath can heal. Scabs eventually fall off on their own, revealing shiny new skin underneath.

If you still have the urge to pick a scab, make these edible pretend scabs that are totally safe to pick instead!

~~~~~~

3 (3-OZ [85-G]) PACKAGES FLAVORED GELATIN POWDER, 2 SHADES OF RED AND 1 YELLOW

BAKING TRAY

PARCHMENT PAPER

SPOON

SHREDDED WHEAT CEREAL (OPTIONAL)

TIP: Remember, if you have a real scab, always keep the area clean and dry to reduce the chance of infection. And don't pick it!

Prepare each package of gelatin according to the package instructions and allow the gelatin to cool until it's just starting to set.

Line a baking tray with parchment paper and spoon the semi-set gelatin gently onto it, forming scab shapes. Experiment with combining the colors to make different kinds of scabs.

If you want a nice crusty scab—because they're the best to pick—sprinkle a little of the shredded wheat cereal on top, if desired.

Place the baking tray in the fridge to set.

Once the scabs are fully set, gently peel them off the parchment paper and place them on your arm, leg or anywhere else you want to look scabby.

Go and show your friends your new scabs.

Adult supervision required

BLOOD TEST TIME

Vampires love it, some people are squeamish about it, an average adult has about 10½ pints (5 L) of it and we all need it to stay alive. Blood travels around the body in a huge network of arteries and veins. It transports oxygen and nutrients around the body, helps our body stay at the right temperature, fights infections and removes waste.

Blood is made up of red blood cells and plasma. Red blood cells are shaped like indented flattened discs and contain hemoglobin, which transport oxygen around the body.

The fake blood in recipe 1 will slowly thicken as the gelatin sets, a bit like blood clotting. Recipe 2 isn't quite as clever but still looks good!

FAKE BLOOD RECIPE 1

1 CUP (240 ML) WATER
SAUCEPAN
WOODEN SPOON
2 TSP (6 G) UNFLAVORED GELATIN POWDER
RED FOOD COLORING
SMALL SPOON OR PIPETTE
PARCHMENT PAPER

Heat the water in the saucepan to boiling.

Turn off the heat and stir in the gelatin powder until dissolved. Let cool for 15 minutes.

Stir in a couple of drops of red food coloring. The blood should thicken slowly.

To make scabs, use a small spoon or pipette to put a few dollops of the blood on parchment paper. It will thicken up like a real scab!

FAKE BLOOD RECIPE 2

3 TBSP (45 ML) CORN SYRUP
BOWL
SPOON
1 TSP FLOUR
RED FOOD COLORING
BLUE FOOD COLORING

Pour the corn syrup into the bowl and stir in the flour.

Stir in a few drops of red and blue food coloring until you get the bright red color you want.

TIP: Note that food coloring may stain clothes and furniture.

PLAY DOUGH BRAIN

Did you know the human brain makes up only about 2 percent of a person's weight yet uses 20 percent of the body's energy? It's also 75 percent water, which is why even being slightly dehydrated can affect brain function.

The outer part of the brain is called the cerebrum and is split into two hemispheres by a central fissure. Each hemisphere has four lobes: the frontal lobe, temporal lobe, parietal lobe and occipital lobe. Under the cerebrum is the cerebellum and the brain stem. The lobes, cerebellum and brain stem are the six main areas of the brain. Around the brain is a liquid called cerebrospinal fluid. This liquid and the skull help to protect the brain from injury and potential infections.

This activity uses play dough to make a model brain that you can dissect. If you were to slice into a real brain, it would feel like cutting through soft tofu!

---

6 DIFFERENT COLORS OF PLAY DOUGH

CHALLENGE

Make a gelatin brain using a brain-shaped mold and a package of flavored gelatin. Try adding a little milk instead of just water when making the gelatin to give it a cloudy appearance.

Have fun dissecting and eating the squishy brain.

Using 4 of the play dough colors, roll each one into 2 thick sausage shapes about 8 inches (20 cm) long. You will have 8 long pieces of play dough total: 2 of each color.

Twist each piece of play dough into a lobe of the brain. Use one piece of each color to form each hemisphere of the brain.

Place the hemispheres together to form a brain!

Now you just need to add the cerebellum and the brain stem. Roll 1 play dough color that hasn't been used yet into a ball shape and place this under the brain. This is the cerebellum.

Roll the final play dough color into a thick sausage shape and place it in front of the cerebellum. This is the brain stem.

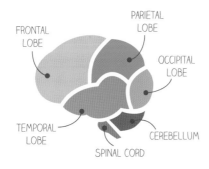

FRONTAL LOBE

PARIETAL LOBE

OCCIPITAL LOBE

TEMPORAL LOBE

CEREBELLUM

SPINAL CORD

RED BLOOD CELL CUPCAKES

Red blood cells, also called erythrocytes, transport oxygen around the body. They collect oxygen from the lungs and are then pumped around the body by the heart via arteries. They then return to the heart via veins, where they are pumped back to the lungs to collect more oxygen.

Red blood cells are biconcave—that is, concave on both sides—and disc shaped, a bit like a donut with no hole.

These red blood cell cupcakes are super gross to look at but taste delicious and demonstrate the unique shape of red blood cells.

RED SUGAR PASTE
RED ICING
CUPCAKES
RASPBERRY SAUCE

Use the sugar paste to make small red blood cell shapes. Form a small sphere shape and then squeeze the top and bottom down at the same time so you have an almost-flat disc with an indent on both sides.

Spread a thin layer of red icing over the top of each cupcake and sprinkle the sugar paste red blood cells over the top.

Pour a little raspberry sauce over the red blood cells and then enjoy the gruesome snack.

FUN FACT

Did you know red blood cells live for about four months? They are made in the bone marrow and contain a protein called hemoglobin that carries oxygen.

BLOOD AND PLASMA COCKTAIL

Did you know that about 7 percent of your body weight is blood? Each tiny drop of blood contains millions of red blood cells and thousands of platelets and white blood cells!

Although blood is a liquid—it has to be so your heart can pump it around your body—it actually contains lots of solid parts. You can discover what these are and how much of them are in your blood by making this gross-looking but totally delicious blood cocktail!

SODA WATER OR LEMONADE
TALL GLASS
POMEGRANATE SEEDS
BANANA
CRUSHED PINEAPPLE
SPOON
STRAW

Blood is about 55 percent plasma, which you're going to represent with soda water, so pour soda water into the glass until it's just over half full.

The other 45 percent of blood is mostly made up of red blood cells, so add the pomegranate seeds until the glass is almost full. There are just a few other things to add.

Less than 1 percent of blood is white bloods cells, so cut a chunk of banana, or another fruit, so it's slightly smaller than the red blood cells, and add that to your cocktail.

Like white blood cells, platelets also make up less than 1 percent of blood volume. Add a tiny amount of crushed pineapple to the mixture to represent platelets.

Give it a stir and drink up!

If you're not feeling well, add a little extra banana since your body makes extra white blood cells to fight infections.

BLOOD FACTS

For every 1 white blood cell, there are around 40 platelets and 600 red blood cells!

Red blood cells transport oxygen from the lungs around your body and remove carbon dioxide.

White blood cells help protect you against diseases and foreign invaders.

Plasma removes waste products and worn-out cells.

Platelets are there to clump together to stop the bleeding if you cut yourself.

SQUISHY BLOOD BATH

A blood bath is another fun way to learn about all the parts that make up blood.

Imagine you could shrink yourself and have a bath in your blood. You could laze around in thick plasma, watching millions of red blood cells carrying oxygen and carbon dioxide float by—perhaps even catching a glimpse of platelets and white blood cells.

Just watch out for the waste products floating by. They might not be very pleasant!

2 (3-OZ [85-G]) PACKAGES UNFLAVORED (NO COLOR) GELATIN
SMALL CONTAINER
RED, WHITE AND YELLOW POM POMS
SPOON
SMALL DOLL (OPTIONAL)

CHALLENGE

Red blood cells are actually shaped like a biconcave disc, which means they are concave on the top and bottom. Try making one with play dough.

Prepare the gelatin according to the package instructions. Add a little extra water so it doesn't set completely. This is the plasma that makes up 55 percent of your blood volume.

Pour the gelatin mixture into the container until it's just over half full.

Add red pom poms into the mixture until the container is almost full and give it a good mix. These are the red blood cells that make up around 45 percent of your blood volume.

Place a couple of white and yellow pom poms into the mixture. These are the platelets and white blood cells, which make up less than 1 percent each of your blood volume. White blood cells are bigger than red blood cells and platelets are smaller than red blood cells.

If you want to really make your concoction a blood bath, add a small doll to the container.

BAFFLING BODIES

Sometimes they smell funny; they produce all kinds of icky, slimy substances; and they need a lot of watering and refueling. The human body is incredible and kind of gross in all sorts of weird and wonderful ways.

Every day your body fights off invaders and removes toxins and waste products from the air you breathe and the food you eat. It can seal itself again after injury thanks to the power of clotting, can grow and repair itself and can pump 10½ pints (5 L) of blood a minute!

In this chapter, you can simulate what your skin looks like close up and then eat it—the model, that is, not your skin. Discover what happens to bones without calcium, dissect a stomach and follow the treacherous journey of food from mouth to rectum.

Find out why urine smells SO bad and create some heathy and not-so-healthy samples to test. Make a stretchy bladder and find out how many of you would fit alongside your intestines if they were stretched out along the ground.

Discover the grossest, the smelliest and the yuckiest features of the human body while marveling at its sheer incredibleness.

Human skin has three main layers: the epidermis, dermis and hypodermis.

Epidermis: This is the outer layer of the skin—the part you can see. It's the body's first layer of defense against bacteria, toxins and other sources of infection. Cells on the outer layer are constantly dying and new cells are replacing them.

Dermis: This layer is thicker than the epidermis and holds blood cells, hair follicles and sweat glands. It also contains collagen and elastin, which are stretchy.

Hypodermis: This layer lies under the dermis and connects the skin to the muscles and bones underneath. It contains subcutaneous fat that helps the body stay warm. Hair follicles also start in this layer.

 Adult supervision required

GELATIN SKIN MODEL

Did you know the skin is an organ? It's part of the integumentary system that also includes your hair, nails and exocrine glands.

Skin is a really important part of the human body. It's the first barrier against germs that can make you ill, helps to regulate body temperature and houses thousands of receptor cells that allow us to feel what something is like when we touch it.

You can be as creative as you like with the materials you use to create your skin model, but the list below is a great starting point.

DID YOU KNOW: Hands, feet and lips have more touch receptors than the rest of the body, making them extra sensitive?

---

2 (3-0Z [85-G]) PACKAGES FLAVORED GELATIN POWDER, 1 YELLOW AND 1 RED OR PINK

2 SMALL SQUARE OR RECTANGLE CONTAINERS

SMALL CINNAMON CANDIES

SPOON

PLATE

LICORICE OR THIN CANDY STRINGS

THIN TORTILLA WRAP

CREEPY FACT

The skin you can see is really dead skin cells! Every minute of every day you lose around 30,000 dead skin cells from the surface of your skin!

Prepare each package of gelatin according to the package instructions. Pour the yellow gelatin into one container and place it in the fridge to set.

Pour the red gelatin into a second container and add the small cinnamon candies to represent red blood cells and sweat glands. Place it in the fridge to set.

Once the yellow gelatin is set, use a spoon to mix it up in the container. This is the layer of fat.

Place the yellow gelatin on the plate so that it resembles the shape of the container it set in.

Add some candy strings to represent blood vessels and nerves.

Carefully place the red gelatin (dermis layer) on top.

Cut the tortilla so it fits on top of the dermis and add some licorice strings to represent hair. Remember, hair actually starts in the hypodermis layer.

DIGESTION MODEL

Have you ever wondered what happens to the food you eat after you swallow it? Imagine a tube longer than a bus. That's how far your food has to travel through the twists and turns of your digestive system.

The process begins in the mouth, where food is broken down into smaller pieces and digestion starts thanks to an enzyme in saliva called amylase. Food then passes down the esophagus and into the stomach, which contains acids to kill bacteria and break the food down further.

Nutrients from the food we eat are absorbed by the small intestine; then the large intestine absorbs water and any nutrients that weren't absorbed by the small intestine. Finally, undigested food becomes feces or poo that is excreted from the body via the rectum and anus.

~~~~~~~~~~~~~~~~~~~~~~~~~~~~~~~~

1 GRAHAM CRACKER

1 SMALL BANANA, PEELED

2 TBSP (30 ML) WATER

2 TBSP (30 ML) ORANGE JUICE

MEDIUM ZIPLOCK BAG

1 LEG FROM A PAIR OF TIGHTS

CONTAINER OR JAR

TRAY OR PLATE

TIP: Note that food coloring may stain clothes and furniture.

Place the graham cracker, banana, water and orange juice into the bag. The water represents saliva and the rest represents food that has been eaten. Make sure there is no air in the bag and seal it.

Squeeze the bag for 20 seconds, crushing up the contents. The food should start to look a bit slimy. This represents the food being ground up in the mouth. Saliva contains digestive enzymes that start to break down the food as the teeth chew it into small pieces. The slimy nature of the food and saliva mixture allows it to pass down the esophagus into the stomach.

Food actually remains in the stomach for several hours as it's mixed with stomach acid that breaks it down further. This gloopy, acidic mixture gets released slowly into the small intestine for the next part of its journey. You don't need to squeeze the food in your ziplock bag stomach for hours; just a few minutes should be enough.

(continued)

DIGESTION MODEL (CONTINUED)

Once the stomach contents feel like a thick liquid, stop squeezing it and put it to one side. Stretch the tights over the top of the container and pour the stomach contents into the tights (small intestine).

Here the pancreas adds digestive enzymes to the mix to break down fats and proteins into molecules that can be absorbed by the body. The liver and gallbladder add bile to dissolve fats. This happens right at the start of the journey through the intestine and neutralizes the acid from the stomach to prevent it from harming the small intestine.

Hold the tights over the container and gently squeeze the liquid out. The liquid in the container represent the nutrients the body absorbs. Keep squeezing until no more liquid comes out.

The food left behind in the tights represents waste products that cannot be absorbed.

Cut a hole in the bottom of the tights and squeeze the contents onto the tray. This is the poo!

CHALLENGE

Change the food you "digest" in the model and investigate if you can make a different color poo!

Old red blood cells and bile, which is green, are added to the food mixture in the small intestine. Add a few drops of red and green food coloring to the ziplock bag just before emptying it into the tights (technically, the food coloring should be mixed in the tights, which represent the small intestine, but it's just much easier to mix food coloring in a plastic bag than in tights) and see how they change the color of the end product. You'll find it is a much darker brown color, similar to poo!

STRONG BONES

Bones are strong because they contain lots of calcium—specifically calcium carbonate. As humans get older, they lose calcium from their bones faster than it is replaced. People with very low levels of calcium have bones that are brittle and more likely to break.

This activity uses vinegar to remove the calcium from a chicken bone to show how bones lose strength when they lose calcium. The bone will become so weak that it becomes bendable.

BONE FROM A COOKED CHICKEN, WASHED

JAR OR CONTAINER

VINEGAR

CHALLENGE

Fizzy drinks are also acidic, which is why they are not good for teeth. Try soaking a bone in a fizzy drink and compare the difference to vinegar. Remember to soak the same size bone for the same amount of time to get an accurate comparison.

Hold the chicken bone in both hands and gently try to bend it. It should feel strong and not bend without a lot of force.

Fill the jar with vinegar almost to the top and carefully place the chicken bone in the vinegar so it is completely covered.

Leave the bone for 48 hours, rinse and try to bend it again.

Put the bone back in the vinegar for another 24 to 48 hours and try to bend it again. The bone should bend easily by this point, but if it still feels hard, leave it in fresh vinegar for another 24 hours.

FUN FACTS

Vinegar is an acid. It dissolves the calcium from the chicken bone. Calcium is the substance that makes a bone hard. Without calcium, bones are much softer.

Exercising and eating lots of calcium-rich foods such as milk, cheese and almonds can help prevent bones from becoming brittle as people get older.

Adult supervision required

WHAT'S IN A STOMACH?

You probably eat a meal and then don't really think about it again, but your body works busily for hours and sometimes days digesting food to get the nutrients and energy it needs to run smoothly.

After being chewed in your mouth and mixed with saliva, the food heads down the esophagus and into the stomach, where it is churned around and mashed by the strong muscles in the stomach walls. Stomach acid kills bacteria and enzymes start to break down the food further before it's slowly passed into the small intestine. The liquid, acidic, gloopy mixture made by the stomach is called chyme. If you've ever been sick, you probably found that your vomit tasted disgusting and a bit acidic. This is because your stomach acid partly consists of hydrochloric acid.

Stomach acid is pretty strong, but luckily a protective mucus lining prevents the acid from damaging the stomach walls.

In this activity, you're going to dissect a gelatin stomach to discover what's been eaten.

---

1 (3-OZ [85-G]) PACKAGE OF UNFLAVORED (NO COLOR) GELATIN

BOWL

VARIOUS FOODS (FOR EXAMPLE, SWEET CORN, BEANS, CANDY, PEPPERS, ETC.)

POTATO MASHER

KNIFE AND FORK

CHALLENGE
Create the stomach of an animal with a particular diet. For example, orangutans eat leaves, fruit and sometimes a bit of soil.

Prepare the gelatin according to the package instructions and pour it into a bowl.

Boil the food items until soft and mash them with a masher.

Add the mashed food to the bowl of gelatin and mix it together.

Place the bowl in the fridge to set.

Carefully tip the gelatin stomach contents out of the bowl and use a knife and fork to dissect it to see what you can find.

STRETCHY BLADDER

Urine, pee or wee is how your body gets rid of waste and water that it doesn't need. You've probably noticed that when you drink a lot of water, your pee is almost clear, and when you're dehydrated, or haven't drank enough water, it's a dark yellow color and has a stronger smell.

Kidneys filter waste from the blood and make urine to get rid of it. Without this process, toxins would build up in your body and make you sick.

Urine contains, among other things, water, salts, ammonia, and a waste product of proteins called urea.

Urine is stored in the bladder. The walls of the bladder contain muscle fibers that stretch as it fills with urine. Sensors detect the stretching and signal your brain to tell you to use the bathroom. A human bladder can hold about 1 quart (1 L) of urine, which is incredible when you consider that an empty bladder is about the size of a pear.

In this activity, you're going find out how much a balloon bladder grows in size when a quart (1 L) of water is added.

BALLOON

WATER

1-QUART (1-L) JUG

Stretch the mouth of a balloon over the end of a faucet and add what you think is 1 quart (1 L) of water. You should have a very full balloon bladder

Empty the water into a 1-quart (1-L) jug to see if you added the right amount of water. Did you over or under fill your balloon bladder?

Keep trying until you get close to 1 quart (1 L).

URINE TEST TIME

We all make it, some of us expel it more often than others and it can be an indicator of overall health. Wee or urine comes in all different shades of yellow. Sometimes it's odorless and sometimes super smelly!

Heathy urine is a light-yellow color. Clear urine means you've been drinking a good amount of water. Dark wee can mean you're dehydrated or indicate an infection. If you wee too much, this can just mean you've been drinking a lot or it can be a sign of diabetes.

Next time you go to the toilet, give your wee a sniff—but don't get too close! You'll probably smell ammonia, which is normal. If you've ever been in a bathroom that smells REALLY bad, there was probably old wee on the floor, as the ammonia smell gets stronger the longer it is left.

Remember, if you experience any changes in your weeing habits or are concerned about anything, see a medical professional.

In this activity, you'll make fake urine and ask a friend to try each sample. Can they identify the healthiest urine? It might not be the one that tastes the best!

½ CUP (120 ML) APPLE JUICE

1⅓ CUPS (320 ML) WARM WATER

RED FOOD COLORING OR RED JUICE

1 TSP SUGAR

SPOON

4 CONTAINERS

Container 1: 2 tbsp (30 ml) apple juice

Container 2: 2 tbsp (30 ml) apple juice, 7 tbsp (105 ml) warm water

Container 3: 2 tbsp (30 ml) apple juice, 7 tbsp (105 ml) warm water, 1 drop red food coloring or 1 tsp of red juice

Container 4: 2 tbsp (30 ml) apple juice, 7 tbsp (105 ml) warm water, 1 tsp sugar

Container 1 contains less liquid than the others and is a darker color. This is a sign of dehydration.

Container 2 represents the healthiest wee.

Container 3 should be a slightly red color, which could mean blood in the urine or ingesting too many beets or red food coloring.

Container 4 should taste extra sweet. This can be a sign of diabetes. Did you know doctors used to taste patients' urine for this very reason? GROSS!

Never drink real urine, even if it's yours!

CHALLENGE

Next time you eat asparagus or beets, check the smell and color of your wee. It may change color or have a distinctive smell.

HOW LONG ARE YOUR INTESTINES?

After food has been churned up and mixed together in the stomach, it's released in small amounts into the small intestine.

As food is squeezed along the small intestine, it's broken down into small molecules the body can use. These are absorbed through the small intestine lining into the blood. All along the inside of the small intestine are folds covered in tiny projections called villi. The outside of the small intestine is covered in blood vessels. The villi absorb nutrients and pass them into the blood vessels for transport around the body.

The bits of food that can't be used are passed into the large intestine, where they eventually come out as poo.

It might be called the small intestine, but it's not small at all! If you stretched it out, it would be about 22 feet (6.5 m) long. No wonder it takes food about four hours to journey through it!

The large intestine is shorter—only 5 feet (1.5 m) long—but thicker.

Both intestines are coiled around each other so they fit inside your body.

YARDSTICK
ROLL OF TOILET PAPER
TAPE
LARGE ROLL OF PAPER
PENCIL

First, use a yardstick to measure 22 feet (6.5 m) on the ground. Roll out the toilet paper so it's as long as your small intestine.

Roll the toilet paper so it's like a tube and tape it at various points to secure the tube.

Roll the large roll of paper out on the floor and draw around a friend so their body shape is on the paper.

Draw the heart, lungs and stomach in the correct place.

Coil the toilet roll intestine around so it fits inside the drawing.

CHALLENGE

Work out how many of you and your friends would fit along your small intestines if they were laid out in a straight line.

EARWAX SNACKS

It's sticky, shiny and sometimes a bit annoying, but what exactly is earwax?

Earwax is made in special glands in the skin of the outer ear canal. If everything is working properly, the wax slowly makes its way to the opening of the ear, where it falls out every now and then. Gross!

Earwax keeps the ear canal moist and contains chemicals that help protect the eardrum and inner ear from infection. It also traps dirt, dust and other things you don't want inside your ears.

LARGE SAUCEPAN
4 APPLES, PEELED AND CHOPPED
1 CUP (240 ML) WATER
WOODEN SPOON
MINI MARSHMALLOWS
TOOTHPICK
CINNAMON

TIP: We've used a pretend cotton swab here, but remember not to use a real one to poke around your ear, as the ear area is very delicate and you could end up pushing the wax further in.

In a large saucepan, heat the apples and water until the mixture has a thick, sludgy consistency.

Leave to cool.

Place a marshmallow on each end of the toothpick.

Sprinkle cinnamon into the apple mix to make it a darker earwax color.

Dip the ends of the marshmallow toothpick into the apple mix.

You now have a cotton swab full of earwax.

STINKY SMELLS

Have you ever walked into a room and smelled an odor that reminded you of a certain place or event? This is because the brain links smells with memories. How clever is that!

If something smells, it gives off tiny particles that float through the air up into the nose. From there, tiny cells called olfactory neurons send messages to a part of the brain called the olfactory bulb. Our brain then works to make sense of the messages so we know what we are smelling. It's not just handy for knowing when to come down for breakfast; your sense of smell can actually save your life by detecting smells, such as certain gases, that might be dangerous.

In this chapter, you're going to discover which smells can move across a room the fastest, if your breath smells and how to make your wee even smellier!

Find out which animal burps three bathtubs worth of air EACH DAY and what to do with a stinky egg!

SMELLY BREATH

Everyone gets bad breath sometimes, but why does it happen? The medical name for bad breath is halitosis. It can be caused by many things, including infections, eating certain foods and poor oral hygiene.

If you don't clean your teeth properly, bits of food get stuck in between your teeth, causing bacteria to build up, and that can smell too!

Common foods that cause bad breath are garlic, onions and some spices. Foods like apples, melons, cinnamon and ginger all help keep your breath smelling fresh and clean!

The next time your mouth tastes a bit funny or people start to move away when you get close, try one of these strategies for finding out if your breath smells. Keep in mind that these techniques won't work if you've just cleaned your teeth.

HOW TO TELL IF YOUR BREATH SMELLS

TECHNIQUE 1

Cup your hands over your mouth and nose. Squeeze your fingers together so no air can escape.

Breathe out slowly and then inhale the breath through your nose.

If the air smells, your breath might not be the freshest. You'd better floss and brush your teeth.

TECHNIQUE 2

Lick the back of your hand and let the saliva dry.

Once it's dry, smell the area where you licked. If it smells, chances are your breath does too!

TECHNIQUE 3

Ask a good friend to smell your breath. Maybe you can smell each other's breath?

TIP: Occasional bad breath is perfectly normal, so don't worry if you don't smell minty fresh all the time. However, halitosis could be a more serious infection, so do seek medical advice if keeping your mouth clean doesn't make you less stinky.

CHALLENGE

Try techniques 1 and 2 as soon as you wake up in the morning, both after cleaning your teeth and after eating. Record the smelliness on a scale of 1 to 10, with 1 being fresh and clean and 10 being the stinkiest of stinky breath.

Brushing your teeth and flossing helps get rid of trapped food that might lead to nasty smells.

STOP THE SMELL

Things that smell give off tiny molecules that float through the air and into your nose.

Imagine little smell molecules drifting around a room. It's a nice thought if they're coming from a freshly baked cake, but less pleasant if they originated in a bathroom.

Once inside the nose, the molecules attach to smell sensors that send signals to the brain, and your brain very cleverly works out what the smell is.

5 DIFFERENT FOODS WITH A STRONG SMELL OR FLAVORED ESSENCE (E.G., VANILLA EXTRACT, PEPPERMINT EXTRACT)

5 COTTON BALLS

5 PAPER CUPS

TINFOIL

TIMER

PEN

PAPER

Place either a smelly food item or two to three drops of essence sprinkled onto a cotton ball into each cup.

Cover each cup with tinfoil so the smell can't escape.

Ask a helper to stand at one end of the room holding a timer while you stand at the other end with the paper cups.

Remove the tinfoil from one paper cup as the helper starts the timer.

The helper needs to stop the timer when they can smell the contents of the cup. Move closer if the smells aren't reaching your helper.

Record the time for each food to discover which smell travels the fastest!

CHALLENGE
Try some less appealing smells. Perhaps garlic, boiled eggs or onion?

HOLD YOUR NOSE

If you've ever had a bad cold, you may have noticed that your blocked nose affected your sense of taste.

There's a lot more to what we taste than what goes on in our mouth. How the food looks and especially how it smells are also big factors.

How food smells can also be a warning. Have you ever opened a loaf of bread and noticed it smelled a bit funny? Or opened the fridge to find a nasty odor coming from moldy or spoiled food?

One very simple way to test how much smell influences what you taste is to hold your nose while eating.

STRONG TASTING FOODS SUCH AS ORANGES OR OTHER CITRUS FRUITS
DIFFERENT FLAVORED JELLYBEANS OR SWEETS
BLINDFOLD

Hold your nose and try one of the foods. Does it taste like it usually does?

Now grab a helper and blindfold them. Make sure they don't have any food allergies.

Instruct them to cover their nose, then give them a jelly bean. Can they tell what flavor it is?

Another way to do this is to puree the food you offer them so they can't tell what it is from the shape or texture.

DID YOU KNOW?

When you eat, chemicals are released from the food and travel up your nose. It's these chemicals, in addition to your taste buds, that tell your brain about the taste.

When you have a blocked nose, the food chemicals can't reach the receptors in your nose, which is why you don't experience the full flavor of the food. So the next time you've got a cold and something doesn't taste as marvelous as it looks, blame your blocked nose!

EGGS AND FARTS

If you've ever packed a boiled egg in your lunch or made an egg sandwich, you've probably noticed the smell. Maybe people made a funny face as you opened your lunch bag and the smell reached them.

Several things happen when you boil eggs. The yolk releases iron, and the white releases hydrogen and sulfur. The familiar eggy smell is from hydrogen, sulfur and iron reacting together to create a smelly compound called hydrogen sulfide.

Hydrogen sulfide is also what makes farts smell bad!

If an egg smells bad—worse than just the usual boiled egg smell—it might have spoiled, so DO NOT eat it! A gray or green yolk is a sign that the egg has been cooked too long, but it will still taste good.

This activity is just for fun but makes boiled eggs a bit gross looking. Remember to notice how they smell as you peel off the shells!

2 EGGS
SMALL SAUCEPAN
WATER
2 ZIPLOCK BAGS
FOOD COLORING

Place the eggs in a small saucepan and cover them with water.

Bring the water to a boil and simmer for about 5 minutes.

Carefully remove the eggs to cool.

Once the eggs are cool, bang them on a hard surface so the shells crack but don't fall off.

Place each egg in a separate ziplock bag and cover with food coloring. Make sure the whole egg is covered and leave for 15 minutes.

Carefully remove the shells to reveal your colored eggs. The food coloring has seeped into the eggs where the shells have cracked, giving them a marbled effect.

CHALLENGE

Boiled eggs will stay fresh in the fridge for a few days, but if you want to keep them for longer than that, you'll have to preserve them. Pickling is one way to do this. Simply boil the eggs, remove the shells, place the eggs in a sterilized jar and cover them with vinegar. Leave them for a few days before eating them.

COW BURPS

Burping, belching, whatever you call it—we all do it, even tiny babies. People often burp when they've swallowed extra air from eating too quickly or drinking something fizzy.

Cows and other animals that graze on grass tend to burp more than humans because of the microbes in their stomachs that help break down the tough cellulose in the grasses. These microbes produce methane gas, which is expelled from cows at both ends.

Cows burp and fart somewhere between 40 and 80 gallons (150 and 300 L) of methane per day. The average bath uses about 25 gallons (100 L) of water, so cows burp about three bathtubs worth of methane each day! There are a LOT of cows in the world, so that's an awful lot of gas being belched out on a daily basis.

This activity shows you how to make a burping bag! It's a little messy, so it's a good idea to do this either outside or on a wipeable surface.

1 PAPER TOWEL

3 TBSP (40 G) BAKING POWDER

½ CUP (120 ML) WATER

FOOD COLORING (OPTIONAL)

½ CUP (120 ML) VINEGAR

MEDIUM ZIPLOCK BAG

Lay the paper towel on a flat surface and place the baking powder in the middle.

Pour the water, a few drops of food coloring (if using) and vinegar into the ziplock bag.

Fold the paper towel over the baking powder and carefully place it inside the bag.

Quickly seal the bag, give it a quick shake and put it on the ground.

The bag will fill with carbon dioxide from the reaction between the baking soda and vinegar. Eventually the pressure of the gas inside the bag will make it burst open!

Note that this is just a pretend burp. Real burps don't involve baking soda or vinegar.

WHAT'S THAT SMELL?

Asparagus makes it smell, beets make it red and a lot of B vitamins can make it bright yellow! The color and smell of urine is affected by what we eat and drink.

Asparagus is one food that is commonly known to make urine smell a bit funny, but did you know not everyone can smell it and some people even like it? The source of the smell is thought to be asparagusic acid being broken down into sulfur compounds. Sulfur is also the compound that makes farts and eggs smell so bad.

Healthy urine doesn't generally have a strong smell, but stinky wee is usually a sign of dehydration.

PAPER

PEN

BOILED ASPARAGUS

PUFFED WHEAT CEREAL

Spend a couple of days recording how your wee smells and looks in the morning, at lunch and before bed. Wee usually smells more strongly in the morning when it's more concentrated. Don't get too close to the wee. Just take a peek in the toilet when you're done. The smells should reach your nose without you sticking your head in the toilet!

Eat some asparagus with your lunch and take note of how your wee smells. There should be a distinct odor a couple hours after you eat it—unless you're one of the people who can't detect the smell.

The next day, eat a bowl of puffed wheat cereal for breakfast. After that your wee should have a malty smell, sort of similar to the cereal itself!

CHALLENGE
Next time you eat asparagus, try cutting off the tips to see if that stops the smell.

JOKING
AROUND

Trick your friends with fake vomit, poo and even an edible cup of dirt with these hilariously gross science activities that aren't for the squeamish.

Make a dung beetle bath bomb for an unsuspecting recipient or give the gift of frogspawn slime.

Pop some pretend pimples and blindfold friends before asking them to put their hands in pots of slimy spaghetti and bug slime. Who'll shriek the loudest?

Just remember to never eat real dirt, and it's generally best to stay away from vomit and poo if you can.

Real frogspawn definitely needs to be left where it is, but you can squish, squeeze and poke the pretend version as much as you like.

I'm pretty sure no one has ever wished they were a dung beetle, but you've got to admire their ingenuity and ability to ignore their stinky home, which doesn't smell anywhere near as nice as the dung beetle bath bomb you'll make.

WHERE'S THE VOMIT?

If you've got a nasty stomach bug, feel very nervous, get car sick or have perhaps been on a roller coaster or spun around too much, you might find yourself vomiting, throwing up, puking, being sick—whatever you want to call it.

Vomit is basically partially digested food that gets pushed up out of your stomach and into your mouth. It usually tastes bad, but how bad depends on how long it's been in your stomach mixing with stomach acid.

It will taste extra nasty if some has come up from the intestines, as it might contain bile, which will make it look a bit green too.

Throwing up is the body's way of getting rid of something bad—unless you've just been spinning around too much, in which case it's probably just motion sickness.

You can make pretend vomit using gelatin and mashed up food. It's just gross enough to trick a friend, but not so gross that it'll make you vomit for real.

~~~~~~~~~~~~~~~~~~~~~~~~~~~~~~~~

SAUCEPAN

WATER

VEGETABLES (LEFTOVERS IF POSSIBLE)

POTATO MASHER

1 (3-OZ [85-G]) PACKAGE OF YELLOW
GELATIN POWDER

9 × 9" (23 × 23-CM) SQUARE PAN

Boil the vegetables on the stove until soft. Drain and leave them to cool.

Mash them up a little with the potato masher.

Prepare about a third of the package of gelatin according to the package instructions. Transfer the gelatin mixture to the square pan.

Add the vegetables to the gelatin once it has started to cool but hasn't set. Place the pan in the refrigerator to set.

Once it's set, mash it all up a bit so it looks like vomit and leave it for an unsuspecting friend to find. Or serve it up as a disgusting snack.

POP THE PIMPLE

Pimples are the most common skin condition in the world. They are perfectly normal, and most people get them at some point in their life.

Under your skin are glands that make oil to keep your skin healthy. Sometimes the oil comes up through tiny holes, called pores, in your skin. The oil traps dead cells and bacteria and hardens to form a plug, which blocks the pore. This creates the perfect environment for bacteria to grow, creating inflammation and swelling. White blood cells swoop in to destroy the bacteria, creating pus as they work, leading to an often very sore pimple!

CREAM CHEESE, SOFTENED
SPOON
SMALL BOWL
ZIPLOCK BAG
SCISSORS
WHOLE PITTED OLIVES OR CHERRY TOMATOES

Stir the softened cream cheese in a small bowl until smooth.

Transfer the cream cheese to a ziplock bag and use scissors to snip off one of the bottom corners of the bag.

Squeeze the bag to pipe the cream cheese into the olives.

Now squeeze the pretend pimples to release the pus!

CHALLENGE

Make a pimple cupcake. Scoop out the center of a cupcake and fill the hole with soft icing. This is the pus.

Spread more icing over the top of the cupcake and add a thin layer of fondant on top.

Make a small hole in the fondant over the pus. Squeeze the top of the cupcake to release the pus!

TIPS: Remember to never scratch, squeeze or pick a real pimple! Though it's tempting (not unlike the scabs from earlier in the book), picking a pimple can make it much worse and possibly lead to infection.

To reduce your chances of getting pimples, wash your skin with soap and water in the morning and evening.

WHAT'S IN THE POT?

Touch is one of the five senses. The sense of touch allows us to determine the temperature and texture of things. Our skin contains millions of touch receptors, but our fingertips, lips and toes contain the most and are especially sensitive.

This game is sure to make an unsuspecting friend shriek. Set up several containers of icky, squelchy substances, blindfold a friend and ask them to guess what grossness they're touching.

UNCOOKED SPAGHETTI

FOOD COLORING

OLIVE OIL

2 SMALL BOWLS

¼ CUP (80 G) CHIA SEEDS

1 CUP (240 ML) WATER

1 TBSP (9 G) CORNSTARCH

WATER

1 (1-OZ [30-G]) PACKAGE UNFLAVORED (NO COLOR) GELATIN POWDER

SPOON

4 SMALL CONTAINERS

FAKE BUGS

BLINDFOLD

SLIMY SPAGHETTI

Prepare the spaghetti according to the package instructions but add a little food coloring to the water. Once cooked, drain and stir in a little oil. Allow to cool.

CHIA SEED SLIME

In a small bowl, combine the chia seeds with 1 cup (240 ml) of water and leave for 24 hours.

CORNSTARCH

In another small bowl, place the cornstarch and slowly add water until it forms a thick paste.

GELATIN

Prepare the unflavored gelatin according to the package instructions and refrigerate until set completely. Use a spoon to chop the gelatin into small pieces.

Transfer each mixture into a small container and add fake bugs to one of them.

Place a blindfold on your unsuspecting volunteer and ask them to place their hands into each container and guess the contents!

CHALLENGE

Put one container in the fridge to chill before starting. Does the cold lead to a bigger shriek?

FAKE POO

Humans tend to avoid it, some animals live it in and some animals even eat it, but animal poo is actually useful to everyone. As it breaks down, nutrients are added to the soil that plants then use to grow.

You can make fake poo to trick your friends or give them a quiz to see if they can identify different kinds of animal poo!

BAKING TRAY
ALUMINUM FOIL
BOWL
WOODEN SPOON
1 CUP (125 G) FLOUR
½ CUP (135 G) SALT
½ CUP (120 ML) WARM WATER
BROWN FOOD COLORING OR PAINT

TIP: If you find real poo, remember to never touch it. Poo can contain harmful bacteria that could make you ill.

Preheat the oven to 300°F (150°C). Line a baking tray with aluminum foil and set aside.

In a bowl, mix the flour, salt, water and food coloring or paint together to make a smooth dough.

Mold the dough into different poo shapes.

Place the poo shapes on the prepared baking tray and bake for 2 hours or until hard.

Show off your poo collection, but don't be like a gorilla and throw it to ward off intruders!

MORE ABOUT ANIMAL POO

One blue whale poo would fill a swimming pool. Imagine a pool of poo!

Dung beetles lay their eggs in poo, which the baby dung beetles then eat!

The color of poo is influenced by the diet of the animal that made it. Blue whale poo is orange because they eat lots of krill, which contain a red chemical.

BUG BATH BOMB

How would you like to live in a ball of someone else's poo? Some dung beetles do! There are three different types of these beetles:

Rollers roll the poo into a ball shape and roll it away. They then either bury the ball to eat later or lay their eggs in it! Imagine starting life in a ball of another animal's poo!

Tunnelers dive into the dung pile and tunnel under it. The female stays in the tunnel while the male brings down dung.

Dwellers live in the dung, also laying eggs in the dung, which provides the newborn baby dung beetles with plenty of food as they grow.

2 CUPS (440 G) BAKING SODA

1 CUP (160 G) CREAM OF TARTAR

1 TO 2 TBSP (15 TO 30 ML) OLIVE OIL

FEW DROPS ESSENTIAL OIL (OPTIONAL)

FOOD COLORING

SPRAY BOTTLE

WATER

SPOON

PLASTIC TOY BUGS

SPHERE MOLDS OR ICE CUBE TRAY

TIP: Note that food coloring may stain clothes and furniture.

Mix the baking soda, cream of tartar, olive oil, essential oil (if using) and food coloring. The mixture should still look very powdery at this point.

Spray the powder 2 to 3 times with water. It will start to fizz. This is just the baking soda reacting with the water.

Keep mixing with a spoon until the mixture starts to feel a bit more solid.

When you get to the point where the mixture can hold its shape and not crumble, you are ready to add it to your molds.

Add your bugs to the baking soda mixture at this point and use a spoon to add the mixture to the molds.

Leave to dry for 2 days.

Drop one in your bath and watch out for the bugs!

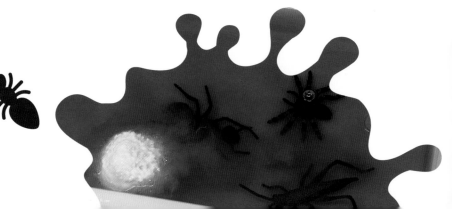

EAT DIRT

First, we're not going to eat real dirt. As wonderful as soil is, it's definitely not a good thing for humans to ingest.

Soil is a mix of decayed plants, animals, rocks and minerals formed over a long time. Lots of factors affect how soil forms, such as the organisms living in it, the climate and the mineral and rocks that are in it. Soil can be sandy, claylike or silty.

You might think of soil as something that just makes you dirty, gets down under your fingernails and covers your knees, but it is actually very important for supporting life on our planet.

Plants suck nutrients up from the soil and use it to hold themselves in place. Without roots in the soil, they would fall over.

Soil helps filter water and recycle nutrients, and it's home to animals, fungi and bacteria. There can be several hundred million bacteria in one teaspoon of soil!

In this activity, you're going to trick your friends into thinking you're eating dirt when actually it's a delicious cupcake topped with crumbled cookies.

CUPCAKES
SILICONE PLANT POTS
CHOCOLATE ICING
CHOCOLATE SANDWICH COOKIES, CRUSHED
SPOONS
CANDY WORMS (OPTIONAL)

Place each cupcake inside a silicone plant pot.

Cover each with icing and top with the crushed cookies.

Add a spoon to each plant pot and a candy worm, if desired.

Eat your dirt!

FROGSPAWN

You've probably seen slimy frogspawn floating in ponds. Frogspawn can contain thousands of frog eggs, but most will not survive to become an adult frog. Being a tadpole is dangerous business—you can dry out before hatching or become a tasty meal for a nearby predator.

Tadpoles transforming into frogs or toads is the most incredible process and also, depending on how you look at it, kind of gross.

Tadpoles take around fourteen weeks to transform into tiny frogs, while toads take longer.

First, the tadpoles eat the jelly around themselves until they are ready to hatch; then they develop a head and tail, followed by gills to get oxygen from the water. Tadpoles are fully fledged froglets when they have front legs that allow them to climb around. Toward the end of the process, the froglet's tail shrinks, lungs develop and back legs grow.

Incredibly, tadpoles can transform faster if they need to escape from a pond full of predators, or if it's too cold, they can delay transformation. How clever is that?

Chia seed slime looks a little like frogspawn and can be used to trick unsuspecting people.

½ CUP (80 G) CHIA SEEDS
BOWL
2 CUPS (480 ML) WATER

Place the chia seeds into a bowl and pour the water over the top. Leave for 24 hours or until you have a thick frogspawn-like gloop!

TIP: Chia seeds are small and crunchy when dry. They taste great sprinkled on cereal and are full of fiber, omega-3 fatty acids, proteins and minerals. However, once you put them in liquid, they can absorb nine times their weight! This turns them from a dry seed into an icky, gelatinous slime!

FUN FACT

Frogspawn is a cluster of jellylike eggs, and toadspawn is more like a long ribbon.

DISGUSTING
DISEASES

Sometimes the human body is pretty gross, especially when it's under attack: oozing snot, pus-filled spots, virus-filled sneezes and swollen, bleeding gums. Luckily, it also has an amazing army of specialized cells ready to jump into action at the first sign of infection.

In this chapter, you will discover how easily germs spread, how to stop them from spreading, the best ways to trap sneeze germs and how to make healthy fake snot and infected green snot. You can even try plucking out damaged marshmallow teeth!

You can also find out why snot is more super than slimy and feast on petri-dish bacteria gelatin.

HOW DO GERMS SPREAD?

There's a reason people tell you to wash your hands before eating. Hands are covered in bacteria. The average person's hands carry around 3,000 different bacteria from more than 100 different species!

When we say germs, we usually mean bacteria or viruses that can make us sick. Germs are everywhere—on your hands, chairs, tables, TV screens and door handles. Kitchen drains and sponges can be dirtier than the toilet!

But not all bacteria will make you sick. In fact, some bacteria are helpful. We have lots of good bacteria in our gut that helps us digest food and prevents harmful bacteria from taking over.

Healthy people can fight off most germs found around the home, but it's especially important to clean food surfaces and utensils to stop the spread of harmful germs.

Although you can't see them, your skin is home to hundreds of different kinds of germs. Your skin acts as a barrier to prevent them from entering your body, but eating with your fingers or putting your fingers in your mouth lets disease-causing bacteria enter your body.

This activity demonstrates how bacteria and viruses can spread between people when they shake hands or if one person touches something and then someone else touches the same thing.

HAND LOTION
EDIBLE GLITTER, VARIOUS COLORS

You need a few people for this activity to be most effective. Each person needs to apply hand lotion to the palms of their hands and then sprinkle glitter over the top. Each person should use a different color glitter.

Make sure each hand has a thin coating of glitter.

Have everyone shake hands with each other.

Each person should now have lots of different colors of glitter on their hands.

Imagine that the specks of glitter are germs. If there's a stomach bug in there, it has potentially spread to everyone!

CHALLENGE

Before removing the glitter, split the group into three.

Group 1 should use a wet wipe to clean their hands.

Group 2 should use just water for 10 seconds.

Group 3 should use soap and water for 30 seconds.

Group 3 probably has the cleanest hands!

Soap and water is the best way to remove germs from your hands and help prevent the spread of infection. It's an easy yet effective solution.

STICKY SNOT

Snot is actually a bit of a superstar when it comes to helping protect the inside of the body from harmful pathogens. The sticky snot in your nose helps filter the air you breathe, trapping smoke, dirt, dust and pathogens that might make you sick before they can reach your lungs.

On a less gross note, snot also warms and moisturizes the air we breathe, making it much more pleasant for the lungs.

Snot is usually clear unless you're fighting an infection. Any green color comes from a chemical released by white blood cells when they're busy fighting the bacteria or viruses that are making you feel sick. Ugh!

MEDIUM BOWL
½ CUP (64 G) CORNSTARCH
WATER
WOODEN SPOON
1 TSP GREEN FOOD COLORING
CINNAMON OR PEPPER

In a medium bowl, pour the cornstarch and slowly add water, mixing until you have a thick liquid. Add the green food coloring to make the mixture look more like snot.

Put your hands in the bowl and squeeze the messy liquid. It should turn hard in your hands—a bit like hard snot.

Let the liquid run through your fingers to create runny snot!

Sprinkle a little cinnamon or pepper onto the surface of the snot. Imagine the cinnamon or pepper is dust or a pathogen that you don't want getting into your lungs. The snot has trapped it and prevented it from reaching the inside of your body.

TIP: Note that food coloring may stain clothes and furniture.

TRAP THE SNEEZE

You've probably always been told to cover your mouth when you sneeze, and for good reason. Sneezing can shoot germs over several yards and up to 100 miles per hour (161 kph)! That's an impressive distance and speed.

Sneezing is the body's way of removing something irritating or harmful from inside the nose.

If you sneeze when you have a cold, this means the mucous membranes of the nose are irritated.

SNEEZE TRAPS

TRAP 1

Hands are the first obvious sneeze trap. They're always there, and if you act quickly enough and keep your fingers closed, you can trap a lot of germs. However, it's kind of pointless if you then use your sneeze-covered hands to open a door or make food for someone else. Did you know the cold virus can survive on indoor surfaces for up to seven days?

If you do use your hands, remember to give them a good wash as soon as you can.

TRAP 2

Sneezing into your sleeve or your elbow is more hygienic than using your hands, but the germs can linger on your clothes. If you use your sleeve to trap a sneeze, change your shirt as soon as you can and put it in the laundry to be washed.

TRAP 3

Tissues are a great way to stop germs from spreading, but remember to wash your hands after blowing your nose and throw the tissue away when you're done with it. Definitely don't give it to someone else to use afterward. They might not thank you for a soggy, germ-filled tissue.

SHOOTING SNEEZES

You've probably had lots of colds or maybe even the flu. Colds and the flu are caused by viruses. There are lots of different cold viruses, which is why you can catch a cold more than once.

When a person with a cold sneezes, the particular cold virus they have flies out of their mouth and nose if they don't trap it (see Trap the Sneeze on page 82), landing on surfaces and people in the same area and possibly infecting them too.

Imagine a room full of people in which one or two have a cold. That's a whole lot of flying viruses going around.

In this activity, you'll make a sneeze shooter using a balloon and water to demonstrate how cold viruses spread.

---

3 TBSP (45 ML) WATER
BALLOON (A ROUND ONE WORKS BEST)
FUNNEL

CHALLENGE

If you don't want to get wet, another idea is to make a pom-pom shooter using a paper cup and a balloon.

Simply cut the bottom off the cup. Tie off the end of the balloon and cut off the top half, leaving behind enough balloon skin to fasten around the cup. Tape the balloon in place.

Drop several pom-poms into the cup, pull back the balloon and watch the pom-pom cold viruses shoot around the room.

Carefully pour the water into the balloon using the funnel.

Blow up the balloon and hold the end closed.

Hold the balloon to the side of your head and let go. Watch as the water flies out. How far have the pretend cold viruses spread?

Ask a friend to stand a yard (1 m) away from you. Does the water reach them?

PETRI-DISH BACTERIA

A petri dish is a small round dish with a lid that fits over the top. Scientists add a layer of agar—a jellylike substance that comes from algae—and grow microbes inside them. The lid fits on top in such a way that germs in the air cannot get in and contaminate the sample.

We're not going to grow our own bacteria in this activity, as it needs to be disposed of properly. Instead, we're going to make some pretty gross pretend petri dishes.

PETRI DISHES

1 (1-OZ [30-G]) PACKAGE UNFLAVORED (NO COLOR) GELATIN POWDER

FLAVORED GELATIN POWDER IN SEVERAL DIFFERENT COLORS

SPRINKLES (OPTIONAL)

Wash and dry the petri dishes.

Prepare the unflavored gelatin according to the package instructions.

Allow to cool for a few minutes and pour a thin layer into each petri dish.

Use the different colors of flavored gelatin powder and sprinkles (if using) to create bacteria on the surface.

Serve your gelatin bacteria plates for dessert.

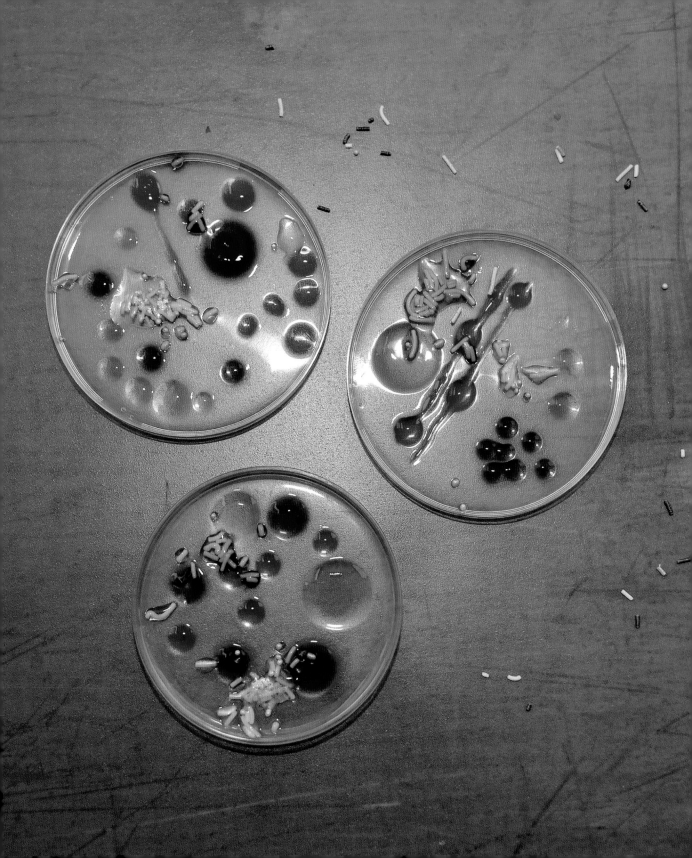

MARSHMALLOW TEETH

Smile! Keeping your teeth clean is super important—and not just when it comes to making them look lovely and white. Unhealthy teeth can be painful and very sore.

Every time you eat, bacteria in your mouth feasts on the sugar. They break it down into acids that can eat away at your tooth enamel—the outer coating of your teeth that protects the more delicate inside. This can leave holes in your teeth called cavities.

Plaque is what we call the sticky, slimy substance that coats your teeth if you don't clean them. Too much plaque can cause gingivitis, which makes gums red, swollen and sore. Plaque is mostly made up of different types of bacteria. It's transparent and hard to see, but if left to harden, it becomes a substance called calculus.

We're actually very lucky to have the modern conveniences of toothbrushes and toothpaste. Years ago, people used chalk and charcoal to clean their teeth. Not quite as minty fresh as toothpaste!

This activity uses peanut butter and jam to make some very dirty, unhealthy-looking teeth.

---

1 COOKIE
STRAWBERRY OR RASPBERRY JAM
WHITE MINI MARSHMALLOWS
PEANUT BUTTER OR HONEY
TWEEZERS

Snap the cookie in half and spread jam around the outer edge of both halves.

Attach marshmallows around the edge of each cookie. These are the teeth.

Spread a little peanut butter—the plaque—on several of the teeth.

Stack the two halves of the cookie on top of each other to make a jaw full of teeth.

Pluck out the dirtiest teeth with tweezers and then enjoy your snack. Just don't think too much about the peanut butter as being plaque and the jam as unhealthy bleeding gums.

After all that sugar, it's probably a good idea to give your teeth an extra good cleaning before bed!

WHY IS SNOT GREEN?

Stretchy snot is the ultimate slime! It's produced by mucous membranes in the nose and serves a very important function: trapping germs, dust and other things you don't want reaching your lungs.

When you're healthy, snot is clear and watery. Green or yellow snot indicates you're fighting a cold. The green color comes from a chemical released by white blood cells when they're busy fighting the bacteria or viruses that are making you feel sick.

If you have more snot than normal, this generally means the body is busy working to get rid of something. Allergies often make a person's nose run because their body is trying to get rid of the irritating allergen. This activity uses psyllium husk to make icky, stretchy snot!

SMALL SAUCEPAN
3 HEAPING TSP (ABOUT 10 G) PSYLLIUM HUSK
1 CUP (240 ML) WATER
WOODEN SPOON

In a small saucepan over medium heat, combine the psyllium husk and water, stirring until the mixture looks like slime.

It should start to stick together very quickly and turn into a clear, stretchy slime after a few minutes! If you find the slime is cloudy looking, add a little more water and keep heating.

SNOT TYPES

Clear – healthy and normal. It is made up of water, proteins, antibodies and salts.

White – a sign of congestion. Swollen nose tissues make mucus flow more slowly and lose moisture.

Yellow – white blood cells have rushed to fight infection.

Green – your immune system is fighting extra hard. The snotty mucus is thick with dead white cells, viruses and bacteria.

CHALLENGE

Create different colors of snot by adding a little food coloring. Note that food coloring may stain clothes and furniture.

GROSS FOODS

If you've ever wondered why foods rot, why we have spit and what happens inside your stomach, this is the chapter for you.

There are rotting pumpkins, an experiment to make your mouth extra saliva filled and even a simulation of the effect of stomach acid on food.

You'll discover why we sometimes crave weird and wacky food combinations and learn how to count your taste buds after turning them a funny color.

Dive in and discover the wonderfully disgusting world of food, taste and digestion.

REVOLTING FOOD COMBINATIONS

Some food combinations just seem to make sense—burgers and fries, mac and cheese, peanut butter and jelly—but some are slightly less appealing.

Strawberries and chocolate? Yes! Pickles and chocolate? Maybe not so much.

What are your thoughts on pineapple and cheese?

There are scientific reasons why some gross food combinations actually work. Foods get their flavor from the chemicals they contain. Foods containing similar compounds have chemicals in common, which makes them taste good together.

BALANCE

Humans tend to like a balance of flavors, so combining foods from several of the five tastes tends to make for a more satisfying dish.

For instance, sweet and salty foods tend to work well together. Imagine maple syrup on bacon or salted caramel. Yum!

HISTORY

Back when food was less plentiful, humans would've craved salt, sugar and fats for survival. This is why we sometimes crave what seem like weird food combinations.

Pickles and ice cream anyone?

THE FIVE TASTES:

Sweet – sweet potatoes, carrots, squashes, parsnips, honey

Sour – vinegar, sourdough bread, lemon and lime juice, yogurt

Bitter – kale, dark chocolate, coffee, green leafy vegetables, olives

Salty – miso, soy sauce, salt

Umami – foods rich in flavor such as meats, tomatoes, cheese, fish

What we call the flavor of a particular food is its unique combination of these five tastes and its smell.

Choose foods from each of the five tastes and create the weirdest food combinations you can. Do they work? Green leafy vegetables with lemon juice sounds yummy . . . but what do you think of dark chocolate dipped in yogurt?

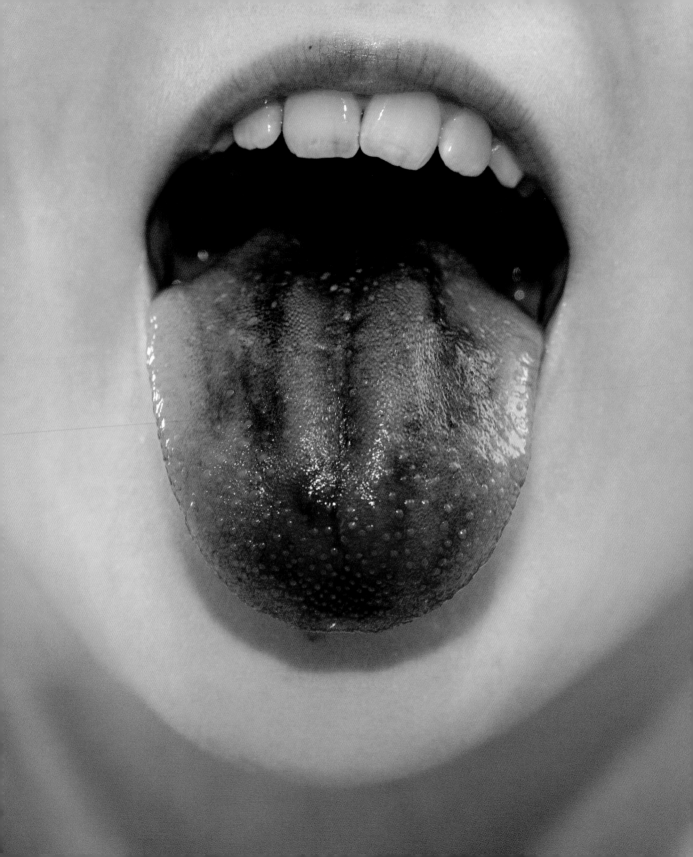

HOW MANY TASTE BUDS DO YOU HAVE?

The average person has around 10,000 taste buds, and they are replaced every two weeks or so. Taste buds are found in the little lumps, called papillae, on the tongue. They have tiny microscopic hairs called microvilli that send messages to the brain about how something tastes. But it's not just taste buds that allow you to taste food.

As food is chewed, chemicals are released that travel up the nose and trigger olfactory receptors. You can thank your nose and tongue for your marvelous sense of taste.

This activity will show you how to count the number of taste buds on a small area of your tongue.

SCISSORS

INDEX CARD

LOLLIPOP OR CANDY THAT CHANGES THE COLOR OF YOUR TONGUE

MIRROR

Cut out a ½-inch (1-cm) square in the corner of an index card.

Suck on the lollipop until your tongue changes color.

Look in a mirror and carefully place the card over your tongue. Count how many taste buds you can see in the square cutout you made in the card.

CHALLENGES

Investigate whether there are different amounts of taste buds on different parts of the tongue.

Try this activity with an older adult. People tend to have fewer taste buds the older they get.

SPIT TESTING

You might think saliva—spit—is a bit gross and just there to keep your mouth moist, but without spit, you wouldn't be able to taste anything. Food has to dissolve in saliva before it can be detected by taste buds.

Saliva is a clear liquid made in your mouth all day, every day. It's made by glands on the inside of each cheek, the bottom of the mouth and under the jaw at the front of the mouth.

Saliva is super useful because it moistens food, making it easier to swallow, and starts the process of digestion. It contains enzymes that start to break the food down before it gets to your stomach.

PAPER TOWEL
DRY FOODS SUCH AS COOKIES, CHIPS AND BREAD
GLASS OF WATER

CHALLENGE

This time, try wet foods such as yogurt and applesauce with a dry tongue. These should be easier to taste than the dry foods as the food chemicals are already dissolved.

Use a paper towel to dry your tongue, and then close your eyes and try one of the dry foods.

How do they taste?

Take a drink of water and try the food again.

Receptors on your taste buds can only detect food chemicals that are dissolved in water or saliva.

If there's no saliva in your mouth, you can't taste the food!

SUPER SALIVA

Roll your tongue around your mouth a bit. Can you feel your saliva? Saliva or spit is the wet stuff that keeps your mouth moist. Salivary glands in your mouth secrete about 1 to 2 quarts (1 to 2 L) of saliva every day!

Saliva is really important because it not only starts to break down food in the mouth and allows you to taste, but it also helps fight infections and cleans the inside of the mouth and teeth. You still need to brush and floss, though. Saliva is super but not a miracle worker.

8 CUPS (2 L) WATER
LARGE CONTAINER
VINEGAR
SMALL CONTAINER
OTHER FOODS

Pour the water into the large container. That's the amount of saliva humans produce each day!

Now take a minute to think about how moist your mouth feels.

Pour a little vinegar into the small container and take a good sniff. This should trigger your mouth to make more saliva.

Smell your other sample foods one by one to see if they make your saliva production increase.

CHALLENGE
Investigate whether smelling foods you like the taste of makes you create more saliva than when you smell foods you don't like.

STOMACH ACID

Stomach acid, also called gastric acid, is another awesome body fluid. It's not just a medium for food to float around in; it helps digest food and kills off most bacteria and microorganisms that manage to sneak into your body. However, while stomach acid is a pretty great bacteria killer, it doesn't destroy everything, so don't use it as an excuse to not wash your hands.

Stomach acid consists of potassium chloride, sodium chloride and hydrochloric acid. It has a pH of between 1 and 3, which is pretty strong. The most acidic pH is 0, and a pH of 7 is neutral. If your vomit has ever been particularly acidic and vile tasting, it probably contained stomach acid.

The stomach walls are coated in a thick mucus layer that protects them from gastric acid.

This activity tests to see how well different acids break down certain types of foods.

MASKING TAPE

MARKER

9 SMALL CONTAINERS

WHITE VINEGAR

LEMON JUICE

WATER

3 IDENTICAL GUMMY CANDIES

3 IDENTICAL SMALL PIECES OF BREAD

3 IDENTICAL SMALL PIECES OF MEAT, COOKED

With the masking tape and marker, label three of the containers vinegar, three lemon juice and three water.

Fill each container halfway with the liquid they are labeled with.

Place one of each food sample into each type of liquid. For example, 1 piece of gummy candy in water, 1 in vinegar and 1 in lemon juice.

Observe how the foods change every 24 hours for 5 days.

Some foods break down faster than others. Foods that contain a lot of sugar are broken down quickly, whereas whole foods high in protein are broken down more slowly. You should find that the meat breaks down the most slowly.

CHALLENGE
Try different types of food and record how they change over time.

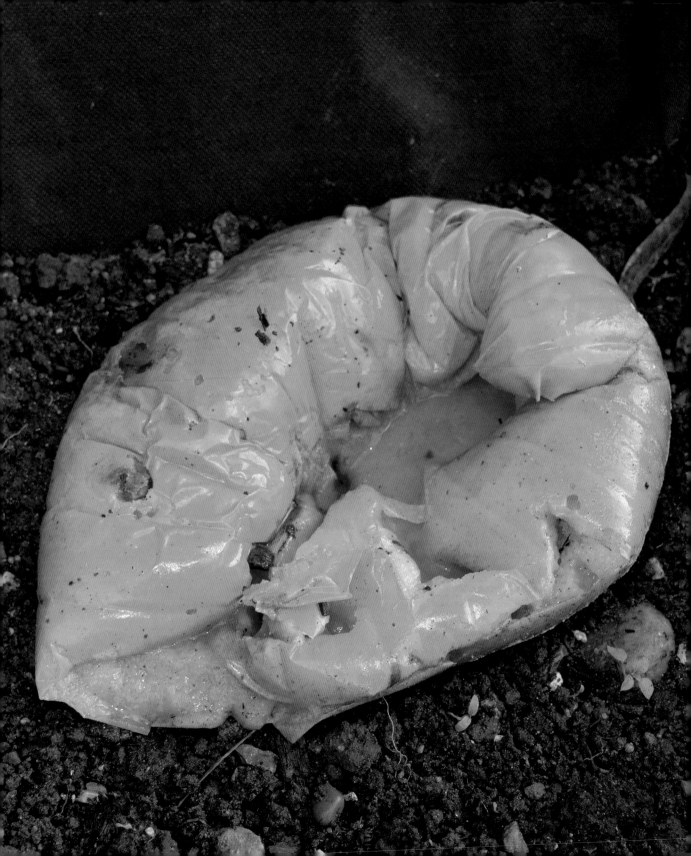

ROTTING PUMPKINS

What do you do with your pumpkins after Halloween? Do you leave them out for wildlife to eat? Do you throw them away? How about using them for a super gross science experiment?

All you need is a pumpkin, carved or uncarved. If it's carved and had a candle burning inside it, remove the candle and any wax.

Leave the pumpkin outside. Somewhere high up off the ground is best so small animals can't reach it.

Check the pumpkin every day and watch how it changes. Once the pumpkin reaches maximum ick level, dispose of it carefully.

TIP: Food starts to rot or decay mostly because of microorganisms that break down the food.

CHALLENGE

Instead of watching it rot, you could try preserving it instead. One idea is to keep the pumpkin cold, since cold temperatures slow the growth of microorganisms. This is why we keep food in the fridge.

PICKLED APPLE SKULLS

Pickling foods in vinegar is a great way to preserve them, so if you find yourself with too many apples, these pickled apple skulls are a great way to prevent them from going to waste. The skulls look gross but taste delicious.

The acid from the vinegar slows down the decaying process by preventing the growth of microorganisms that would normally make the apples rot.

LARGE SAUCEPAN
1 CUP (240 ML) APPLE CIDER VINEGAR
¼ CUP (60 ML) MAPLE SYRUP
¼ CUP (50 G) BROWN SUGAR
½ TSP SALT
1 CUP (240 ML) WATER
1 CINNAMON STICK
5 APPLES, CORED AND SLICED
KNIFE
STRAW
LARGE GLASS JAR

In a large saucepan over medium heat, combine the vinegar, maple syrup, brown sugar, salt, water and cinnamon stick. Heat until the sugar dissolves.

Carefully cut the apples into a skull shape and use a straw to make holes for the eyes and mouth.

Carefully pour the liquid mixture into a jar, add the apple slices and keep in the fridge. They should be ready to eat the next day and last a week in the fridge.

CHALLENGE

Try storing the apple slices in orange juice instead. The orange juice will prevent the apple slices from turning brown by shielding the fruit from exposure to oxygen. It's the oxygen reacting with the surface of the fruit under the skin that makes it turn brown.

ALL ABOUT THE POO

Number two, poo, poop, feces—whatever you call it, everyone does it. Horses and dogs like to sniff it, most humans do it in the toilet, animals do it in all kinds of different places and it all smells.

Human poo is actually about 75 percent water, and the rest is undigested food, waste and bacteria—good and bad. It doesn't sound quite so bad when you put it like that, does it?

You can often identify an animal that left poo behind by studying the poo itself. How the poo looks and smells depends on the food the animal ate. Bird poo is very different from dog poo, for example, because they have very different diets. However, it's not always that obvious. The Australian citrus swallowtail caterpillar is cunningly disguised to look like bird poo, which helps protect it from being eaten.

In this chapter, you can make edible brown and green poo, discover what your poo would be like if you were a sloth, shoot poo like a caterpillar and find out how fast food moves through your own digestive system.

Adult supervision required

BROWN OR GREEN

Poo is brown mostly thanks to yellow/green bile (the alkaline stuff your liver produces that breaks down fats in digested food) and old red blood cells. You know how old blood turns brown? Well, the same thing happens to red blood cells—they turn brown as they start to break down, making your poo brown too!

Yellow poo can be a sign that you've eaten a lot of fat. Green poo usually means you've eaten a lot of veggies or the poo has passed through you a bit quickly. Or it could be a sign of too much food coloring. Too many beets can also turn your poo and wee red!

MEDIUM MICROWAVE-SAFE BOWL

7 OZ (200 G) CHOCOLATE, BROKEN INTO CHUNKS

3 TBSP (45 ML) CORN SYRUP

1 CUP (100 G) CRUSHED SHORTBREAD COOKIES

3 TBSP (50 G) CARAMEL OR FUDGE

3 TBSP (25 G) RAISINS

WOODEN SPOON

PARCHMENT PAPER

In a medium microwave-safe bowl, melt the chocolate in the microwave.

Add the corn syrup, crushed cookies, caramel and raisins to the bowl and mix well.

Use your hands to roll the chocolate mixture into a sausage shape and wrap it in parchment paper.

Leave to cool in the fridge.

Once cool and hard, you can eat the poo! It should look gross but taste amazing.

CHALLENGE

Make green poo by using white chocolate and a few drops of green food coloring.

Note that food coloring may stain clothes and furniture.

TIP: Always see a doctor if you're concerned about the color of your poo. Yellow poo or blood in your poo can be a sign of a serious disease.

WHOSE POO?

Poo is the waste that remains after the nutrients from food that has been digested are absorbed by the body.

Animals produce poo in a similar way, but it comes in different shapes, sizes, smells and even colors! Some poo is so distinctive that you can identify the animal that made it just by looking at the poo. Fish poos are long and thin. Rabbit poo is small, circular and hard. Wombat poo is cube shaped and cow poos are big, flat and smelly!

The poo of meat-eating animals looks very different compared to the poo of plant-eating animals. Meat eaters tend to poo less because meat is easier to digest and there is less waste than plant material. Water is also a factor. Both cows and sheep eat a lot of grass, but cow poo is big and sloppy and sheep poo is dry and small. This is because cows drink a lot more water than sheep.

PLAY DOUGH

Use your hands to mold the play dough into different poo shapes.

Ask a friend to guess which animal's poo you made!

CHALLENGE

Most animal poo is brown or black, but vampire bat poo looks a bit like red jam because they feed on blood. Camel poo is extremely dry since they don't drink much water. Can you model these animals' poo?

SLOTH POO

The life of a sloth seems pretty easy at first glance. They laze around all day, eat, sleep a LOT and move really slowly. The slow habits of a sloth are reflected in their sluggish digestive habits. It can take sloths up to a month to digest a meal compared to about four hours for a human. Now that's sluggish!

Luckily for anyone hanging out underneath a sloth, sloths come down to the ground to poo. They do a poo dance to make a hole for the poo, make the immense poo and then do another dance to cover up the poo. Sounds very civilized, doesn't it? A healthy sloth poo can be a third of the animal's body weight. Can you imagine making a poo that size?

BATHROOM SCALE
CALCULATOR

Weigh yourself on the scale and calculate a third of your body weight by dividing that number by 3.

Find items around the house that weigh the same as a third of your body weight.

That's a BIG poo!

CHALLENGE

The average human poo weighs about 1 pound (450 g), so if humans pooed once a week, their poo would weigh 7 pounds (3.1 kg)! Try to find something that weighs 7 pounds (3.1 kg).

SHOOTING POO

Some caterpillars have the ability to shoot their poo over 4½ feet (1.4 m) away from themselves! Scientists believe they do this to protect themselves from predators and parasites that are attracted by the poop.

They are not the only creatures to do this. Adélie penguins also launch their poo, although not quite as far as caterpillars.

You're not going to try shooting human poo in this activity, because that would be super gross, but you can build a catapult and try flinging some fake poo.

9 WIDE POPSICLE STICKS
4 RUBBER BANDS
MILK BOTTLE TOP
DOUBLE-SIDED TAPE OR GLUE
PING PONG BALL
TAPE MEASURE

Stack 7 popsicle sticks on top of each other. Twist a rubber band around each end to hold them in place.

Place another stick above and one below the stack of 7 so they make a cross shape.

Twist a rubber band around the middle of the cross.

Twist another rubber band around the bottom of the two sticks as you can see in the photo.

Attach a milk bottle top using double-sided tape or strong glue.

Place the ball in the bottle top and experiment with your ping pong ball poo!

Can you shoot it 4½ feet (1.4 m) away?

WHERE DID ALL THE CORN GO?

You might think corn isn't digested at all, since, let's face it, it does looks pretty similar before and after we eat it.

The outer part, the hull, is made up mostly of cellulose, which humans cannot digest. However, the inside is mostly starch, which is easily digested. When you see corn in poo, you're actually only seeing the outer husk. The nutrients from inside the corn are happily being absorbed by your body.

CORN

A NOTEBOOK

A PEN

CHALLENGE

Make play dough poo models that demonstrate the results of digestion that's too fast, just right and too slow.

Too slow – small, hard, look like grapes

Just right – long brown sausage

Too fast – soft and mushy

Add some corn to your play dough poop!

To do the following activity, you'll need to have not eaten corn for a few days.

Once you're sure there's no existing corn in your system, eat some corn! It can be on the cob, frozen or come from a can. Write down the exact day and time you ate it.

Now don't eat any more corn until you've seen the portion you just ate show up in your poo.

HOW LONG SHOULD THE CORN JOURNEY TAKE?

It should take between 24 and 36 hours. If it's less than 12 hours, you're probably not very well, and if it takes more than 36 hours, you might be a bit constipated. Try eating some prunes!

Person 1
Person 2   25.5 hrs
Person 3   30 hrs
Person 4   31.2 hrs

BIRD POO

Bird poo is unmistakable due to its color. While most animals produce brown poop, bird poo is black and white because they don't poo and wee separately. It all comes out in one strange mess. Human wee contains urea dissolved in urine, but birds excrete their nitrogenous waste in the form of a white paste called uric acid.

Bird poo usually has a dark central part—this is the poop—surrounded by white uric acid waste.

In this activity, you're going to make bird poo splats on black paper. Be sure to set up in an area where you can make a mess. Outside might be a good idea.

PAINTBRUSH
WHITE PAINT
BLACK CONSTRUCTION PAPER
BLACK PAINT
CORNSTARCH (OPTIONAL)

Dip the paintbrush in the white paint and flick it over the black paper to make a bird poo splat.

Do the same with the black paint to add some darker spots to your poo splats.

To make the paint thicker, add a teaspoon of cornstarch and mix well.

BUGS
AND GRUBS

Love them or loathe them, bugs and grubs are everywhere, and humans need them! The animals we commonly call bugs are scientifically known as arthropods. Arthropods include insects and spiders.

Worms and slugs are a bit different. They are not arthropods because they don't have an exoskeleton, or a skeleton outside their body, and they don't molt.

Did you know there are around 1.4 billion insects per person on Earth? That's a whole lot of insects. They pollinate flowers, fruits and vegetables and make honey, beeswax and silk.

Insects are often the sole food source for many animals, and some species are even eaten by humans. Life without bugs and grubs would be very different. Worms are important because they help add air to soil and make it more fertile. Likewise, slugs are food for lots of different animals.

In this chapter, you can discover which food ants and other bugs like the most by opening a bug restaurant, set up a home for worms by making your own wormery and make slugs that actually secrete slime!

Adult supervision required

WORM HOTEL

A worm might not be something you look forward to seeing in your garden, but they are super useful. They eat soil, mix it up and release nutrients that plants can use to grow. The tunnels worms create as they move underground also help keep the soil aerated so that air and water pass through more easily.

Worms love the dark; they tend not to come up out of the soil unless it's wet, as they dry out in the sun. This is why if you go for a walk after it's rained, you see more worms than usual. They can also regrow parts of their body, which is either amazingly cool or a bit disgusting depending on how you look at it.

The next time you find a worm looking lost, take it back to your garden. The plants will thank you.

~~~~~~~~~~~~~~~~~~~~~~~~~~~~~~~~

GRAVEL

**OLD JAR OR BOTTLE WITH LID
(PLASTIC IS BEST)**

SOIL

SAND

WATER

WORMS

LEAVES

GRASS CLIPPINGS

BLACK CARDSTOCK OR PAPER

TAPE

TIPS: Keep the worm hotel out of direct sunlight. Worms like the dark.

After it rains is a good time to look for worms because they'll be near the surface. If you can't wait for rain, pour some water on the ground and wait for 30 minutes.

Add a layer of gravel to the bottom of the jar. This should help with drainage inside the worm hotel.

Add the soil and sand in layers. Drop a small amount of water onto the surface, but not too much.

Find some worms and gently put them on top of the soil. Add the leaves and grass clippings on top. Leftover vegetable peelings also work well.

Make some holes in the lid (get an adult to help) and screw it on the jar. Wrap a sheet of black paper or cardstock around the bottle and tape it into place. Keep this over the wormery when you're not looking at it.

You should see the sand and the soil get mixed up as the worms burrow down in the worm hotel. The leaves and grass should also be getting pulled down into the soil.

This is what worms do in the garden. They help mix decaying material into the soil, where it is broken down by microorganisms into nutrients that can be used by plants to grow.

SLIME TRAIL

If you've ever watched a snail or slug move, you'll know they leave a gross slimy trail behind.

Snails are gastropods. Gastro means "stomach" and pod means "foot." So a snail is a stomach foot! Gastropods really are basically a large foot with a mouth at one end. Imagine having a body like that!

Slime is a sort of mucus that helps snails and slugs move. It also helps them stick to whatever surface they happen to be on. Slime is especially handy when they want to climb vertically or even go upside down. Suddenly making slime seems pretty cool, doesn't it? Would you like sticky slime feet?

Slime also helps protect gastropods from hazards such as prickly objects on the floor, bacteria and even sunlight. It might be a bit gross, but it's super useful if you're a slug.

SMALL BOWL
¼ CUP (40 G) CHIA SEEDS
1 CUP (240 ML) WATER
SPOON
SCISSORS
OLD TIGHTS OR MUSLIN FABRIC
SHARPIE OR GOOGLY EYES
GLUE (OPTIONAL)

In a small bowl, soak the chia seeds in the water for about 2 hours. If the seeds on top seem dry, give the mixture a good stir. You need all the seeds to be in contact with the water.

Once the mixture looks thick and gelatinous, it's ready.

Cut out a long rectangle of fabric from the tights and place a spoonful of chia seed slime in the center.

Roll the fabric into a sausage shape and tie the ends. Use a sharpie to create eyes or glue on some googly eyes. This is your slug!

You should find that the chia seed slime seeps out of the fabric, allowing your slug to move easily across surfaces and leave a slime trail!

SUCK BLOOD

Mosquitoes don't actually bite—they suck! A mosquito's mouth is called a proboscis and is actually six tiny needles that find blood vessels. Only female mosquitos bite, and they drink blood to grow their eggs.

Mosquitoes are busiest at night and search for food by detecting body heat and carbon dioxide.

In this activity, you're going to pretend to be a mosquito and suck blood through a layer of skin. Only it's not a real layer of skin—it's just gelatin. And it's not real blood either—it's a smoothie!

THICK RED SMOOTHIE OR PUDDING

SHALLOW CONTAINER

STRAWBERRIES OR RASPBERRIES, SLICED

1 (1-OZ [30-G]) PACKAGE UNFLAVORED (NO COLOR) GELATIN POWDER

STRAW

Pour a ¾-inch (2-cm) layer of the thick red smoothie into the container. Place the sliced strawberries over the top.

Prepare half the gelatin according to the package instructions. Save the remaining gelatin powder for another activity.

Allow the gelatin to cool slightly before carefully pouring a layer of it over the top of the smoothie and then place the container in the fridge to set.

Once it's set, pick up your straw sucker, pierce the gelatin skin and drink the blood.

TIP: The smoothie needs to be very thick so the gelatin will float on top.

HOW MANY WORMS?

Worms are more plentiful in the soil than you probably think.

A square meter of land can contain around 100 worms, depending on where it is. Very hot, cold or dry areas have fewer.

Next time you see worm casts—a fancy word for worm poo—all over your lawn, don't be grossed out. Yes, they have come from inside a worm, but those tiny casts are full of nutrients that help plants grow.

Worms spend most of their time underground, so it's hard to imagine just how many there are slithering around under the soil. This activity shows you how to work out just how many worms are under the ground.

RULER
PIPE CLEANERS
PLAY DOUGH

Measure out a 10-inch (25-cm) square.

Earthworms can grow up to 14 inches (36 cm) but generally are around 3 inches (8 cm) long. Create 25 worms in a variety of different sizes using the pipe cleaners and/or play dough.

Place all your worms in the square.

That's a lot of worms!

CHALLENGE

You can get worms to come to the surface by making the ground vibrate. When worms sense vibrations, they think a predator such as a mole is burrowing down into the soil, so they make their way to the surface to escape. Try banging a spade on the ground and see how many worms appear.

BUG RESTAURANT

What do bugs like to eat? Find out by setting up your very own bug restaurant! The idea behind this activity is to leave several different food items out and watch which insects are attracted to which foods.

~~~~~~

8 PETRI DISHES
ORANGE
MELON OR MANGO
LEMON
APPLE
BREAD CRUMBS
SUGAR WATER
VINEGAR
WATER

Place a similar amount of each food or liquid in each petri dish.

Leave the dishes outside and observe them every hour or so.

Record which foods insects are most attracted to and which insects you see.

You'll probably find that insects were more attracted to some dishes than others.

TIP: You probably found a lot of ants came to try the restaurant. Ants will try a lot of different foods but have a preference for sweet foods. They probably headed toward the orange, apple or mango.

BUG SLIME

Cornstarch is usually used to thicken sauces in cooking, but it also makes a great slime.

Cornstarch is a powder when dry, but when water is added, it flows like a liquid as the solid particles are suspended in the liquid. When you squish it in your hands (apply a force), the particles lock together, making the slime feel solid.

When the force is removed (you open up your hands), the slime goes back to being runny again.

MEDIUM CONTAINER
2 CUPS (256 G) CORNSTARCH
WATER
FOOD COLORING (OPTIONAL)
SPOON
PLASTIC BUGS

TIP: Note that food coloring may stain clothes and furniture.

In a medium container, stir together the cornstarch, water and a couple of drops of food coloring (if using) until you have a thick liquid mixture.

If you scrunch the slime between your fingers, it should feel like a solid ball, but when you open your hands up, the slime should drip through your fingers. If it's too runny, add a bit more cornstarch.

Add the bugs and enjoy the icky slime!

FUN FACT

This kind of slime is sometimes called oobleck. Cornstarch oobleck is made up of molecules arranged in long chains. When the chains are stretched, the liquid will flow, but when you force them together, they stick together to form a solid.

GROSS
HISTORY

If you're not already super grateful to be living in the modern era, you will be after reading this chapter. You'll never look at toilet paper or a clean, sewage-free street in the same way again.

Find out why a famous king's body is said to have exploded, what the Romans used to wipe their bottoms and why many years ago you'd have been lucky to have a leech sucking your blood, as the alternative was probably excruciatingly painful.

Discover what ancient streets smelled like, why people stopped bathing frequently for 300 years and how they tried to mask the disgusting smells in the air, thinking that would stop them from catching the terrifying diseases circulating at the time.

Investigate what happens when vinegar is added to milk and how people not so long ago used the resulting protein to make buttons and ornaments.

Are you brave enough to take this terrifying journey through history?

MAKE A MUMMY

Ancient Egyptians mummified their dead because they believed the body was reunited with its spirit in the afterlife.

The process is pretty gross and not for the faint of heart. First, they washed the body, and then they removed the organs, including the brain. They put the heart back inside the body because they thought it was the center of intelligence and emotion. Then they threw away the brain and put the liver, lungs, intestines and stomach into jars. The body was then filled with salt and left to dry out. After 40 days, the salt was removed and the body was filled with linen before being wrapped in bandages and placed in a coffin called a sarcophagus. The whole process took about 70 days and basically stopped the body from decaying. Decay is mostly caused by bacteria. The salt dried the bodies out so bacteria couldn't survive.

In this activity, you're going to make a vegetable monster and then mummify it!

---

3 SMALL PLATES OR TRAYS
VARIOUS FRESH VEGETABLES
SALT

On each plate, create a small vegetable person.

Cover one with salt, place one in the fridge and leave one out at room temperature.

After a week, check all three plates.

Bacteria don't survive as well at cold temperatures, so the vegetables in the fridge should look healthier than the ones at room temperature.

The mummified (salted) vegetables should have dried out.

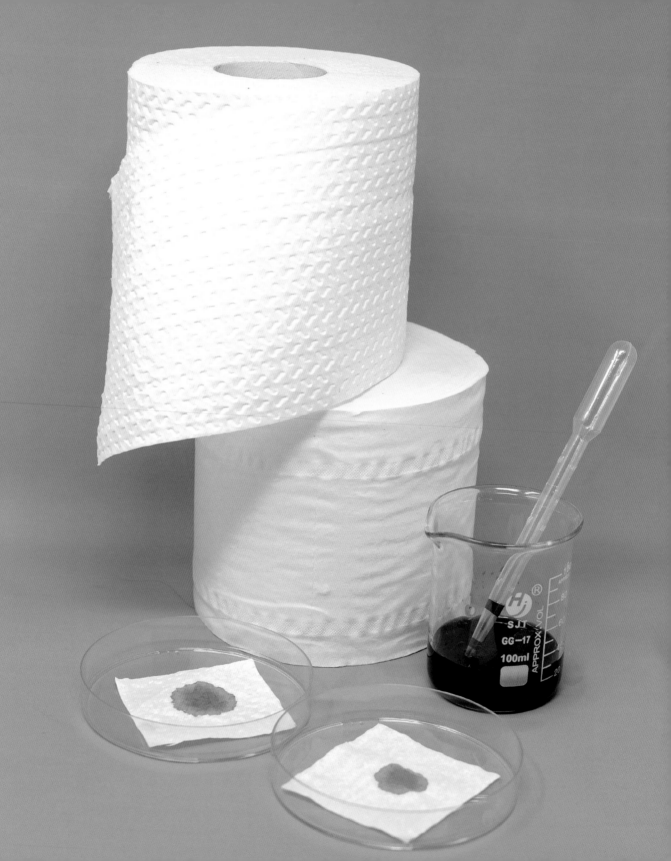

TOILET PAPER TESTING

Ancient Romans used a sponge on the end of a stick, Eskimos used moss or snow, Vikings used wool and Mayans used corn cobs! Luckily for current bottoms everywhere, toilet paper was first invented in China in the fourteenth century, and the familiar roll of toilet paper was invented in 1890.

These days we're lucky to have a huge choice of different types of toilet paper: extra soft, recycled, aloe-infused, luxury and inexpensive. Each piece is only used once and only by one person. It's a world away from a shared sponge on a stick!

What features do you look for in toilet paper? Do you want it to absorb a lot, be soft, tear easily or simply perform the basic function it's needed for?

SCISSORS

A SELECTION OF DIFFERENT TOILET PAPER BRANDS

**PETRI DISHES OR SMALL CONTAINERS
(1 FOR EACH BRAND OF TOILET PAPER USED)**

PIPETTE

WATER

Cut the toilet paper into squares so they fit inside the petri dishes. Each sample should be the same size.

Using the pipette, carefully drop three drops of water onto each sample and observe how well they absorb it. The water will spread out more on the least absorbent paper.

CHALLENGE

Try more toilet paper tests. Investigate the durability of a particular brand by securing a sheet on top of an empty tin can with an elastic band. Sprinkle water on the toilet paper so it's wet and see how many coins the toilet paper can hold before breaking.

Repeat the experiment with other brands of toilet paper. Remember to use the same amount of water to wet each sheet and the same type and number of coins.

Adult supervision required

ORNAMENTS FROM MILK

Have you ever opened a container of yogurt and seen a layer of thin liquid on top? It might look a bit disgusting, but it's actually perfectly normal. Whey is the liquid left behind after milk has been curdled and strained. It sometimes separates from the main yogurt and just needs to be mixed back in.

Milk contains proteins and fats suspended in water. When milk is fresh, the proteins are free to move around. If the pH of the milk drops, the protein molecules stick to each other and create lumps, or curds. The liquid left behind—whey—is watery, green and definitely gross looking. You can make your own whey using milk and vinegar.

SMALL SAUCEPAN

1 CUP (240 ML) 2% MILK

1 TBSP (15 ML) WHITE VINEGAR

WOODEN SPOON

SIEVE

BOWL

PAPER TOWEL

PAINT (OPTIONAL)

In a small saucepan, heat the milk gently until it starts to bubble. Turn off the heat.

Add the white vinegar and stir with the wooden spoon. You should see the solid and liquid parts of the milk separate. Curdling is generally a sign that the milk has spoiled, but curdled milk proteins are actually used to make cheese, yogurt and some desserts.

Use the sieve to separate the curds and whey. Place the curds into a bowl and leave for a few minutes to cool.

The curds can then be molded into shapes and left out to dry. The result is called casein plastic! Just squeeze all the excess water out of the curds and make them into whatever shape you want. Then leave it to dry on a paper towel for about 48 hours. Once it's dry, you can paint and decorate it, if desired.

CHALLENGES

Try adding vinegar to some milk again but without heating it this time. The milk will still curdle but not as quickly as when heated.

Try different edible household acids such as lemon juice and orange juice instead of vinegar.

DID YOU KNOW?

In the early 1900s, milk was often used to make plastic buttons, combs and other items. Plastic is made from molecules called polymers that repeat themselves in a chain, and milk contains casein polymers that allow it to be molded into any shape.

TIP: Add a little food coloring to the wet curds to make colored casein plastic. Note that food coloring may stain clothes and furniture.

THIRSTY LEECHES

Imagine going to the doctor and having your blood drained as treatment. It seems crazy to us now that we know that blood is vital for survival and too much blood loss can actually lead to death.

Bloodletting is thought to be one of the oldest medicinal practices, and it was a common procedure up until the late nineteenth century. Doctors believed that removing blood from a patient would help cure their illness. In medieval Europe, veins or arteries were painfully nicked to allow blood to flow.

If you were lucky, a leech would be used instead of a terrifying surgical instrument. Leeches are worms that feed on blood. They attach themselves to an animal, slice into the skin and suck their blood. Leeches can drink up to ten times their own weight in blood. When they're full, they drop off the skin to enjoy digesting their meal.

SMALL BOWL

WATER

RED FOOD COLORING

CLEAR GUMMY CANDY

In a small bowl, add ½ inch (1 cm) of water and a few drops of food coloring.

Place the gummy candy into the water and leave it for 24 hours.

The gummy candy will expand and turn red as it absorbs the water.

DID YOU KNOW?

When leeches bite, they cleverly inject a substance that stops blood from clotting and numbs the pain so the victim doesn't even realize they've been bitten as the leech sucks away.

Because of the enzyme they secrete that helps blood flow freely, leeches are still used today for certain medical procedures such as those that help skin grafts to heal and restore blood circulation.

SOAPMAKING

Babylonians are thought to have started making soap around 2800 BC. Soap in those times was made from fats and ashes boiled together. Ancient Egyptians mixed animal and vegetable oils with alkaline salts to make their version of soap, and early Romans made soap in the first century from urine! Soap back then was used for cleaning and medicine rather than personal hygiene like it is today.

In the mid-nineteenth century, soaps just for bathing were distinct from laundry soap. As soapmaking techniques improved, soaps became more readily available.

Today soap can be made very simply using a melt-and-pour base with a little essential oil. Ancient Egyptians were some of the first people to use essential oils extensively. They used them for cosmetics, medicine and as part of the embalming process.

MELT-AND-POUR SOAP BASE
MICROWAVE-SAFE BOWL
PLASTIC WRAP
ESSENTIAL OILS
SOAP MOLD

Cut the soap base into small pieces and place them in a microwave-safe bowl.

Cover the bowl with plastic wrap and heat for 30-second intervals until completely melted.

Stir the soap after each minute of heating. The soap will be very hot when it's melted, so take care to ask an adult for help.

Add a few drops of your favorite essential oil to the soap mix and stir.

Carefully pour the scented soap mixture into the mold and allow it to cool.

After a couple of hours it should be hard enough to remove.

Use your soap to get rid of your own nasty odors! Aren't you glad you didn't live in Roman times and had to use urine soap?

MASK THE SMELL

People of the past smelled very differently from how we smell today. Romans used urine to clean clothes, as medicine for animals and for making leather. People gargled with urine, and for several hundred years, beginning in the sixteenth century, people bathed very infrequently—partly because they feared catching diseases in public bathhouses. Soap was in short supply and deodorant was a far cry from what it is today.

Some people tried to mask the smell with expensive perfumes or by wearing small bouquets of flowers attached to their clothes. They thought that breathing in the smell of flowers and herbs would help block out their pungent surroundings—a bit like a modern air freshener!

In this activity, you can make your own mini bouquets of herbs and flowers, called nosegays, and leave them in different places around the house to see how effective they are at masking smells.

FRESH HERBS SUCH AS LAVENDER, ROSEMARY, SAGE, THYME AND FENNEL

STRING OR TWINE

Choose a selection of herbs you think will be best at masking nasty odors and tie them together with the string or twine.

Leave the mini bouquets in the smelliest parts of the house. Do they mask the stinky smells?

CHALLENGE

Create a plague bag of your own by wrapping a selection of herbs in a muslin cloth or tissue and tying them with a string.

DID YOU KNOW?

People believed that the plague in the seventeenth century was caused by the bad smells in the air and carried around plague bags filled with herbs and nice smelling plants. They thought the good smells would stop them from becoming infected by the disgusting air that was tainted with the smells of human and animal waste and rotting food.

EXPLODING BODIES

William the Conqueror was king of England in the eleventh century. His bowels are said to have exploded as his body was being put into his sarcophagus, or stone coffin, filling the abbey with such a smell that everyone fled.

This is thought to have occurred because the bacteria in his gut produced gases that caused pressure to build up inside his body. William's body had to be transported for his funeral and was then delayed. The combination of heat and bacteria meant his body bloated more than normal. Eventually the pressure built up so much that his bowels exploded.

It's not known how true the tale is, but there probably wasn't an explosion. Perhaps it was more of a fart. Today the only remains of William left is a thigh bone in his final resting place.

One way to demonstrate an explosion after a buildup of pressure is using a bottle of fizzy soda and Mentos. Imagine the Mentos are the bacteria and the bottle is a body. Drop the bacteria as fast as you can into the body. The bacteria will produce a lot of gas and—boom!—you have an explosion!

5 TO 6 MENTOS

2-LITER BOTTLE OF SODA

CHALLENGE
Try this activity with both diet and regular sodas to investigate if it makes a difference to the explosion that occurs.

TIP: Do this activity outside.

This activity works best if you have a helper. As one person opens the bottle, the other needs to drop the Mentos in and stand back straight away!

The Mentos need to be dropped into the bottle of soda as soon as it's opened, before too much gas escapes from the bottle.

ACKNOWLEDGMENTS

Where to start? Firstly, I need to give a huge thank-you to my husband and children who were immensely understanding as I wrote this book. There were many late nights, half-finished projects scattered around the house and endless photographs to pose for.

The kids learned not to eat anything I left in the kitchen that looked like part of an experiment, especially if it looked like poo. It's hard to distinguish between chocolate and salt dough! Gelatin wasn't safe either. Sometimes it was pretend vomit or the inside of a stomach. Sometimes it was just dessert.

There were lots of smiles, many ick faces and generally lots of "WHAT are you doing NOW?" faces that amused me greatly and kept me going.

It's never easy to write a book, and it does eat into family and friend time, so thank you everyone for understanding!

Thank you also to all the children who helped out with photos. You were all amazing and made the BEST faces without me even having to ask.

Finally, huge thanks to the lovely Page Street Publishing team for inspiring me to write a third book and once again guiding me through the process so expertly and patiently!

ABOUT THE AUTHOR

Emma is a professional science blogger and author following her dream of making science fun for even the smallest of children.

She has been creating exciting and easy science investigations for Science Sparks since 2011 and can't think of anything she would rather be doing.

Emma is a mum to four children whose endless questioning and creativity inspire many of her ideas.

Emma is also the author of *This Is Rocket Science* and *Snackable Science Experiments*, both bursting with easy and fun science experiments and activities for kids of all ages.

INDEX